The Tour de France

Solving Addition Problems Involving Regrouping

Joseph A. Saviola

PowerMath™

Published in 2004 by The Rosen Publishing Group, Inc.
29 East 21st Street, New York, NY 10010

Book Design: Michael J. Flynn

Photo Credits: Cover © AFP/Corbis; pp. 5, 9, 14, 21 © Mike Powell/Getty Images; p. 5 (flag) © Eyewire; pp. 6, 13, 18 © Ezra Shaw/Getty Images; p. 10 © Doug Pensinger/Allsport; p. 17 © Werner Dieterich/ The Image Bank; p. 23 © Pascal Rondeau/Allsport.

Library of Congress Cataloging-in-Publication Data

Saviola, Joseph A.
The Tour de France : solving addition problems involving regrouping / Joseph A. Saviola.
p. cm. — (PowerMath)
Summary: Gives facts about the Tour de France bicycle race, such as how far participating cyclists travel each day, and shows how to use addition to determine further information.
ISBN 0-8239-8963-1 (Library binding)
ISBN 0-8239-8851-1 (Paperback)
6-pack ISBN: 0-8239-7324-7
1. Addition—Juvenile literature. 2. Problem solving—Juvenile literature. [1. Addition. 2. Problem solving. 3. Tour de France (Bicycle race)] I. Title. II. Series.
QA115.S38 2004
513.2'11—dc21

2002154484

Manufactured in the United States of America

Contents

The Tour de France

The Tour de France is the most famous bicycle race in the world. Each year, the best **cyclists** from around the world take part in the race. The Tour de France happens every July in France and lasts for about three weeks. During that time, the cyclists cover a distance of over 3,000 **kilometers** (2,000 miles). It is a trip that takes them over flat land and up steep mountains. The **route** changes every year.

Most of the Tour de France takes place in France, but the race's route sometimes takes cyclists into neighboring countries like Italy, Spain, Belgium, and Germany.

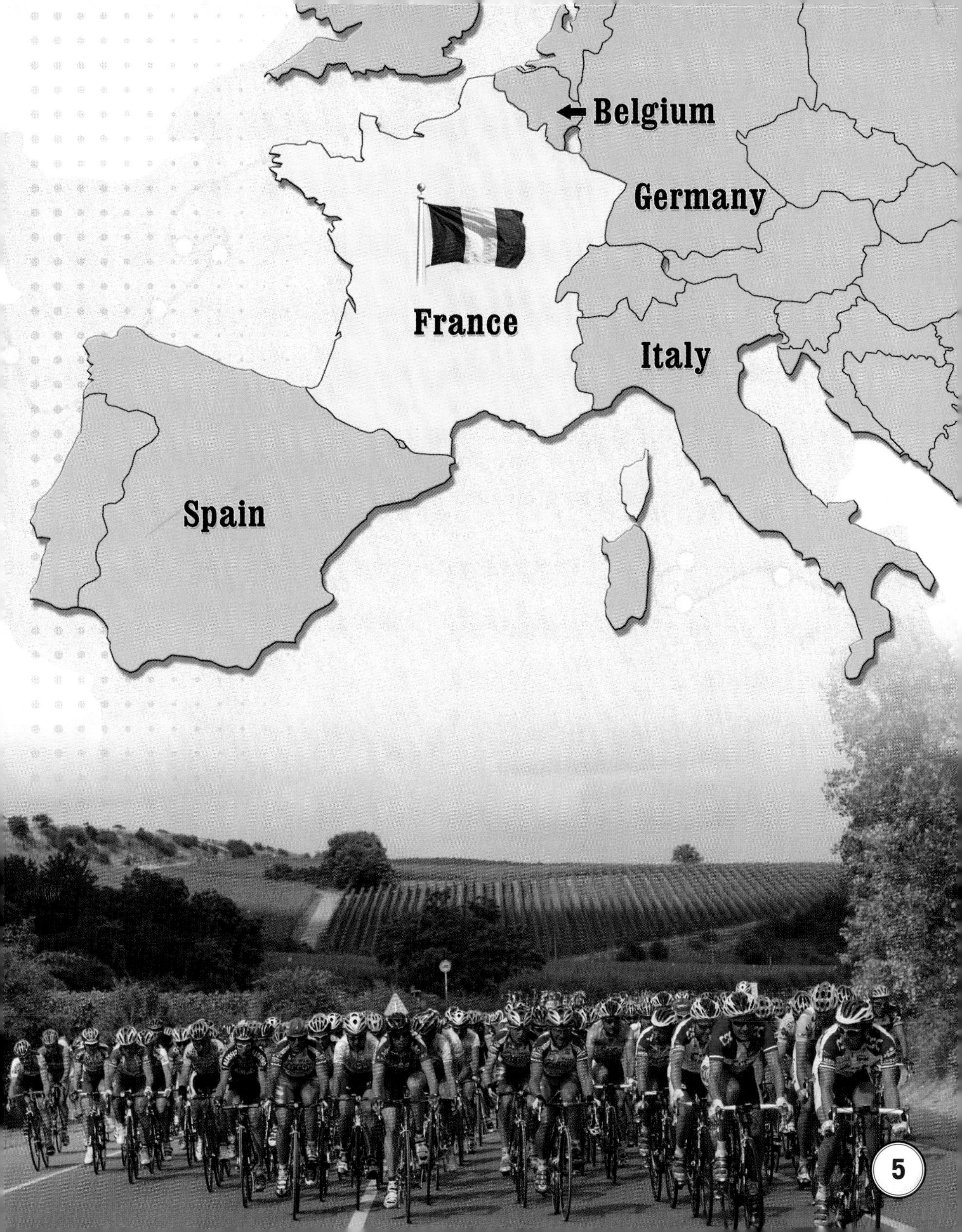
Belgium
Germany
France
Italy
Spain

The Tour's History

The Tour de France was started in 1903 by a French cyclist and journalist named Henri Desgrange, who ran a French newspaper. The race gained popularity in 1910, when the riders rode a **dangerous** route through the Pyrenees (PEER-uh-neez) Mountains in southwestern France. This route was called the "circle of death."

In 1919, the race's famous yellow **jersey** was introduced. The yellow jersey is a shirt that is worn by the cyclist who has the shortest total race time at the end of each day.

American cyclist Lance Armstrong is shown here wearing the yellow jersey that identifies him as the leader after completing a stage of the race.

Teaming Up for the Tour

The cyclists in the Tour de France belong to teams. There are usually about 20 teams that take part in the race. Each team is made up of 9 cyclists.

Although only one cyclist can win the race, the fastest cyclist on each team could not win without the help of his teammates. If someone has a flat tire, a teammate might give him one of their wheels. Teammates might also ride in front of the leader to protect him from the wind, or might block an **opponent** who is trying to gain time.

When a cyclist wins the Tour de France, all of the winner's teammates share the prize.

If there are 20 teams in the race and each team has 9 cyclists, how many cyclists ride in the race? You can multiply 20 by 9 to get your answer!

$$\begin{array}{rl} 20 & \textbf{teams} \\ \times\ 9 & \textbf{cyclists per team} \\ \hline 180 & \textbf{cyclists} \end{array}$$

When you add up a column and the total is more than 10, write down the number in the ones place, then carry the number in the tens place to the next column. Follow the same steps for carrying numbers to the hundreds place.

	1 1	
First Day	**8**	**kilometers**
Stage 1	**160**	**kilometers**
Stage 2	**+ 195**	**kilometers**
	363	**kilometers**

Starting the Race

Let's take a look at a sample race route to see how far the cyclists must ride to finish the race. In this sample route, the cyclists must start by finishing a short ride of 8 kilometers in Paris, France. Paris is the largest city in France. It is also France's capital.

The next day is the first full day of riding. Each full day of riding is called a "stage." During Stage 1, the cyclists must cover a distance of 160 kilometers. The following day—Stage 2—the cyclists must ride 195 kilometers! How far do the cyclists ride in these first three days?

This Tour de France route starts at the Eiffel Tower, a famous landmark in Paris, France. The Eiffel Tower was built in 1889. It is 984 feet tall!

The Cyclists Race On

So far the cyclists have ridden 363 kilometers, but their journey has just started! During Stage 3, they must ride 160 kilometers. During Stage 4, they must ride another 68 kilometers. For Stage 5, they have to ride 196 kilometers. So far, Stage 5 is the longest distance they have covered in one day. Let's see how far they have traveled during this part of the race.

The Tour de France is France's biggest yearly sporting event. Huge crowds watch the race in person, and millions of people around the world watch it on TV.

	2 1	
Stage 3	**160**	**kilometers**
Stage 4	**68**	**kilometers**
Stage 5	**+ 196**	**kilometers**
	424	**kilometers**

Do the Math

How far have the cyclists gone by the end of Stage 5? You can add 363 kilometers to 424 kilometers to get your answer!

363	**kilometers**
+ 424	**kilometers**
787	**kilometers**

Stage 6	**230**	**kilometers**
Stage 7	**226.5**	**kilometers**
Stage 8	**+ 211**	**kilometers**
	667.5	**kilometers**

Do the Math

How far have the cyclists gone by the end of Stage 8? You can add 787 kilometers and 667.5 kilometers to get your answer.

	1 1	
	787	**kilometers**
+	**667.5**	**kilometers**
	1,454.5	**kilometers**

Into the Mountains

By the end of Stage 5, the cyclists have gone 787 kilometers. During Stage 6, they ride 230 kilometers and finish the day in Lyon (lee-YOHN), the third largest city in France. For Stage 7, the cyclists travel 226.5 kilometers. For Stage 8, they must ride another 211 kilometers, ending their day in the French Alps in southeastern France, close to the border of Italy. The Alps form a mountain chain that stretches 660 miles across Europe.

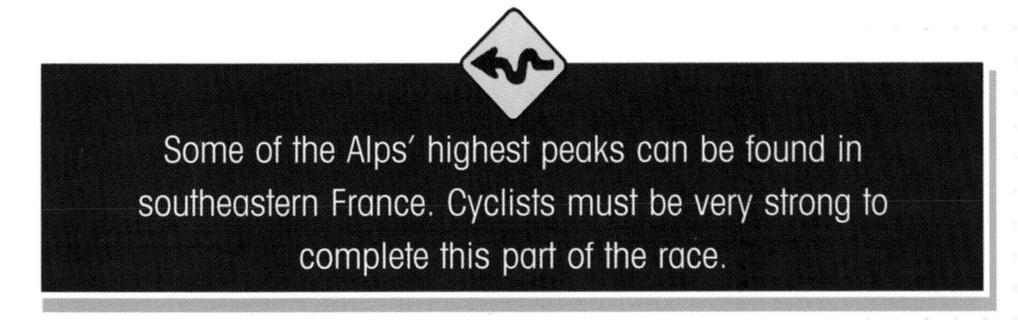

Some of the Alps' highest peaks can be found in southeastern France. Cyclists must be very strong to complete this part of the race.

The Race Continues

So far, the cyclists have gone 1,454.5 kilometers. For Stage 9, they ride 184.5 kilometers. During Stage 10, they ride 195 kilometers to Marseille (mar-SAY) on the southern coast of France. Marseille is France's second largest city.

The cyclists are finally given a day to rest. They are then taken to their next starting location, a small city west of Marseille. For Stage 11, they ride another 160 kilometers.

Marseille lies on the Mediterranean Sea.
It is France's largest seaport.

	2	
Stage 9	**184.5**	**kilometers**
Stage 10	**195**	**kilometers**
Stage 11	**+ 160**	**kilometers**
	539.5	**kilometers**

Do the Math

How far have the cyclists gone by the end of Stage 11? You can add 1,454.5 kilometers and 539.5 kilometers to get your answer.

11	
1,454.5	**kilometers**
+ 539.5	**kilometers**
1,994.0	**kilometers**

	2 1 1	
Stage 12	**48.5**	**kilometers**
Stage 13	**197.5**	**kilometers**
Stage 14	**+ 191.5**	**kilometers**
	437.5	**kilometers**

1 11	
1,994	**kilometers**
+ 437.5	**kilometers**
2,431.5	**kilometers**

Do the Math

How far have the cyclists gone by the end of Stage 14? You can add 1,994 kilometers and 437.5 kilometers to get your answer.

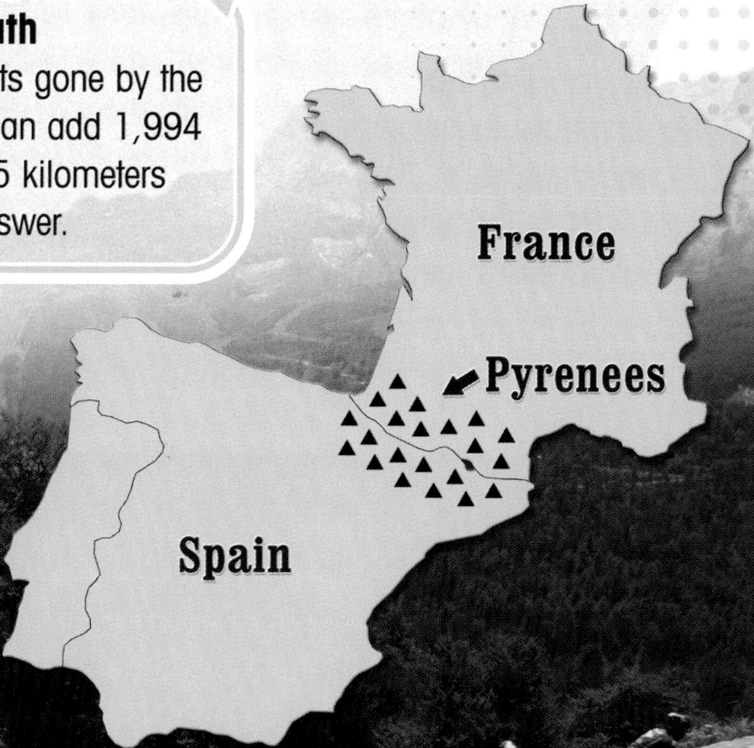

France

Pyrenees

Spain

The Pyrenees

The cyclists have gone 1,994 kilometers so far. That's about 1,200 miles! For Stage 12, they must ride 48.5 kilometers. During Stage 13, they must ride another 197.5 kilometers. They must then ride 191.5 kilometers to complete Stage 14, ending their day in the Pyrenees Mountains of southwestern France. The Pyrenees Mountains make up the border between France and Spain.

Some cyclists ride more quickly than others on flat land. Other cyclists are better at riding up steep mountains. The cyclist who wins the Tour de France usually does very well in both areas.

More and More Mountains

So far, the cyclists have traveled 2,431.5 kilometers after riding for just over two weeks! For Stage 15, they must ride 159.5 kilometers. The following day the cyclists get to rest. During Stage 16, the cyclists travel 197.5 kilometers. These two stages are very difficult, taking the cyclists through the Pyrenees Mountains. Then the cyclists must ride another 165 kilometers to finish Stage 17 in a city called Bordeaux (bor-DOH) in southwestern France.

	2 2 1	
Stage 15	**159.5**	**kilometers**
Stage 16	**197.5**	**kilometers**
Stage 17 +	**165**	**kilometers**
	522.0	**kilometers**

Do the Math

How far have the cyclists gone by the end of Stage 17? You can add 2,431.5 kilometers and 522 kilometers to get your answer.

	2,431.5	**kilometers**
+	**522**	**kilometers**
	2,953.5	**kilometers**

They Made It!

By now, the cyclists have gone a total of 2,953.5 kilometers. They are entering the final stages of the race. Stage 18 covers a distance of 200 kilometers. During Stage 19, the cyclists ride another 49 kilometers. For Stage 20, the last stage, the cyclists ride 160 kilometers to end the race in Paris, where the Tour de France began! The cyclist with the lowest total time for all stages of the race wins!

	1	
Stage 18	**200**	**kilometers**
Stage 19	**49**	**kilometers**
Stage 20	**+ 160**	**kilometers**
	409	**kilometers**

Do the Math

How far have the cyclists gone by the end of the race? You can add 2,953.5 kilometers and 409 kilometers to get your answer.

1 1	
2,953.5	**kilometers**
+ 409	**kilometers**
3,362.5	**kilometers**

Glossary

cyclist (SY-kluhst) A person who rides a bicycle.

dangerous (DANE-juh-rus) Something that is not safe.

jersey (JUR-zee) A special kind of shirt often worn by people who take part in sports.

kilometer (kuh-LAH-muh-tuhr) A unit of length equal to 1,000 meters or .62 mile.

opponent (uh-POH-nuhnt) Someone who is on the other side in a contest.

route (ROOT) The path that a person travels to get from one place to another.

Index